A MOTORCYCLE COP'S MOTORCYCLE MANUAL

Fifty years of accumulated motorcycle lore…

And a few good stories at no extra charge!

Kenneth James Crist

A Motorcycle Cop's Motorcycle Manual

BY KENNETH JAMES CRIST

Hekate Publishing First Edition, 2019

ISBN: 978-1-912017-87-4

Hekate Publishing

73 John Drive

Farmingville, NY 11738

Cover Image by A. F. Knott

admin@hekatepublishing.com

https://www.hekatepublishing.com

Dedication

This little manual is dedicated to every cop in Wichita who ever "rode the wheel" and kept motorists safer, whether they liked it or not. Special thanks to Lt. Jim Stansbury and Lt. Charles Lauffer for keeping us all alive…

TABLE OF CONTENTS

INTRODUCTION

When I began this project, I felt there was a definite need for some kind of little, compact book a new rider could carry and use as a reference to get through that dangerous and sometimes painful start-up phase of motorcycling, and I thought, who better to fill that need than myself: a retired motorcycle officer and a fifty-year owner and rider of bikes? Who else could better cut through the usual hype and bull crap and get down to what folks really want to know about motorcycles and the things that go into making a good, safe rider?

As I set to work, I envisioned a nice, small pamphlet with a few do's and don'ts and an anecdote or two just to keep it light. Then I began finding more material that "really should be included". What you are holding in your hand as you try to decide if it's worth your time and money is a project that got way beyond what I had envisioned.

Chapter after chapter unfolded and, as is sometimes the case with my fiction, the dang thing just sort of wrote itself.

I hope when you finish this, you'll feel you got your money's worth and I hope you'll think enough of it to pass it on to someone else. If it saves one life or one limb, it's worth it. Trust me.

THE MOTORCYCLE OFFICER *(WHO THE HELL ARE THOSE GUYS?) WHY THEY DO WHAT THEY DO-AND HOW TO AVOID MEETING THEM.*

"Why would anybody ever want to be a cop?" Just like everyone else who ever embarked on a law enforcement career, I got a lot of that when I started into the Wichita Police Academy. It was the 70's, and in those days, I had a standard answer. "I just wanna ride motors."

"You wanna be a *motor cop*? Jesus, man, you really *are* crazy!"

Well, maybe that's what it takes. I had been in law enforcement and security since I was seventeen, (U.S. Air Force Security Police, Beech Aircraft Security) and I decided to quit fooling around and go for the real thing. At the time, there was no department in the country more respected than the Wichita Police Department.

Their motorcycle officers rode big, gleaming white Harleys and that was for me, by God. I had applied every year for five years and, just like clockwork, every year I got turned down. "Yer too little, kid! Gotta be five-nine and a hundred-fifty. Then, in 1972, they lowered the height requirement by two inches and the weight requirement by ten pounds and suddenly, I was hired.

The academy was no breeze, fourteen weeks in the high summer of Kansas and the physical training about killed me, but I made it through and went to the streets for patrol duty. As soon as I had my required two years in uniform patrol, my plan was to transfer to traffic division and await the next opening for a motorcycle assignment.

As my two-year goal approached, I was getting both salty and savvy. I was becoming a pretty good street cop. I walked into the squad room one morning and noticed several motor officers looking dejected and pissed off at the world.

"What's up, guys?" I asked, and they dropped the startling bombshell: The motorcycle division was being disbanded. The bikes would be sold, except for a very few for the ceremonial squad to ride for funerals and parades. The officers were to be reassigned to other divisions or to other assignments within the traffic division.

It would be eight years before a different Chief of Police would decide that the change was a mistake and bring them back. By that time, I was thirty-eight years old and when I got my shot, I was the oldest officer in the school. I was still the oldest ten years later when I retired with twenty years service—ten of it on motorcycles. I had survived ten years riding the "wheel" with only one minor spill and one reportable accident—when an elderly gentleman backed into my motorcycle when it was parked, cracking the front fender.

So what makes a motor officer "tick"? Why do they go out in the worst traffic day after day and take the risks? I can guarantee it's not the pay. On the Wichita Police Department the motorcycle assignment didn't even pay extra. There was no hazard pay and shortly after I retired, they stopped the officers from taking their assigned machines home with them. Then, another new Police Chief and, suddenly, the motorcycle division was gone again. Now, yet another Chief, this time an import from Minnesota, and the department has purchased eight new Harleys. We'll see what happens next…

I can also tell you officers don't take up riding bikes because of prestige. It might seem like a plum of an assignment for a while, but after several months or years it becomes a job, just like anything else.

But the motorcycle officer serves a need in his community and it's often a need only he or she can fill. The officer on the bike is often the only one who can get to the scene of the wreck or whatever's snarling up traffic. The fact that the motorcycle can "split" between lanes of traffic or go down the sidewalk, (or even up and down stairs, for that matter) gives the bike cop an advantage over the guys in cars. The power and responsiveness of his bike often means he can be the first on a scene or the lead officer to run interference for the ambulance that's coming to help your loved one.

But much of the motor officer's time is devoted to the duty that most street cops hate—traffic enforcement. When I rode for the City of Wichita, I was proficient with six different radar machines, plus two breath analysis machines to measure levels of breath alcohol. The typical motor officer makes about 2,500 to 3,500 vehicle stops every year and tickets the major portion of those drivers. Taking into consideration that of all the law enforcement officers who are gunned down each year in our country, 36% are shot on so-called routine car stops, the danger level for a motor cop goes up proportionately.

Did I ever feel I was in danger? Lots of times. Sometimes, several times a day. Did I ever find people with guns? Damn straight I did. And dope. And booze. And anything else that was illegal.

Did I ever let anyone go? Sure, all the time. But most people's excuses when they're faced with the motor cop and his radar gun are so feeble and overworked, the veteran cop doesn't even take the time to listen. He's quite literally heard them all a thousand times: "Officer, I was just keeping up with traffic." *Could be true, could be an outrageous lie—doesn't matter to me, doesn't matter to a judge—you were speeding, and I handle violators one at a time. I know I can't catch 'em all, but buddy, I caught you!*

"My speedometer must be off." *Get it checked. I check my radar every time I write a ticket. I know it's accurate. Your speedometer is not my problem, but your speeding is. Have a nice day!*

"Officer, that light was yellow." *The light*

was redder than a spanked baby's ass and you already know this. I saw the look on your face when you ran it and then saw me watching you. I saw you shake your head and say, "Aw Shit!" when I pulled out into traffic. I'm prepared to testify as to exactly where your car was when the signal changed and how fast you were going. Hey, Mr. Clueless, here's a tip for ya: If you hadn't been speeding and yakking on your cell phone, you'd have had plenty of time to stop for the light.

And on it goes, an endless litany of excuses for poor driving habits and bad judgment. Do excuses ever work? Well, once in a while somebody will come up with one that's just so good ya hafta let 'em go. An example of this happened to me when I was less than a year from retirement.

Routine car stop—violation, speed. Young white male driver, alone. As I pull out the ticket book he asks, "Officer, if I can give you an excuse you've never heard before, will you cut me some slack this time?"

"I might," I reply, "but I'll warn you, I've been doing this a long time. I doubt if you've got anything to say that I haven't already heard."

"Well," he began, "I talked to my wife a

while ago from the office and she assured me she was going to get pregnant this afternoon, and I'd sure like to be there for it."

As I handed him back his license with a smile and a warning, I just had to ask. "How many times has that worked for you before?"

"Just once," he replied, "on a Highway Patrol Trooper in western Kansas."

As he drove away, I thought to myself, *"Well, it's always nice to know you're in good company."*

So just what is the attraction? Why do guys (and gals) put themselves in danger on a daily basis for one of the most thankless jobs you could ask for? I think it's a combination of things. First, there's the freedom. In Wichita, motorcycle officers got very few calls and even though they were assigned an area of responsibility, they were not required to *stay* in their area, so they had the run of the entire city.

Then there's the challenge of just staying alive and in one piece. Most motor cops learn to take it in small chunks, but the level of concentration required to survive and look good all day on that white *Kawasaki 1000*

Police Special guarantees you'll be tired at the end of your tour, whether you actually *do* anything or not.

There were days when the weather was perfect and everything was going right and I actually marveled that I was *being paid* to ride a bike all day. There were days when it was miserably cold or unbearably hot and I was low on my ticket count and I wondered if a trip to the department shrink wasn't in order, to get my screws re-threaded. I *had* to be nuts to be doing this job!

A friend on the department may have summed it up best when he said, "Well, Kenny, it's like we're going huntin' or fishin' every day, except there's no bag limit, you can always shoot does and ya never have to throw anything back because it's too small. They're *all* keepers."

I could bore you to tears with stories about my career, but that's not what this book is about. As we move along there will doubtless be more "motor cop lore", but for now it's time to look at *your* situation and see if I can help.

In the chapter heading I spoke about motorcycle officers and I wrote, "…and how to avoid meeting them." When you ride a bike, you really do have an advantage, you know. A motorcycle is just about the hardest target to pick up on a radar gun and the truth is, if there's anything anywhere around you, such as a car or truck, the radar is going to read them and not you. Simple physics. The gun reads the *strongest* return signal. Not the fastest. Whatever is presenting the largest frontal mass, distances being equal. Now, if you're appreciably closer, then you may be in trouble. If the cop's got a laser gun, you're definitely in trouble. The laser beam is so narrow it can pick targets out of the traffic mix, to the exclusion of all others. Doesn't matter if you're the lead or not, or how large your vehicle is. Best not to speed in the first place. But then, we all get the urge now and then, don't we?

So, what about those times when you *do* get stopped? What's the best thing you can do *then*? Here are some do's and don'ts to help you along.

Do: Pull over in a timely fashion, where indicated. Do it safely, render a signal and pay attention to what the nice officer is trying to tell you. Normally, pull to the right-hand side of the roadway and stop. If the officer wants you to pull off onto another street or parking lot, he'll direct you, either with one of those obnoxious overhead speakers or with some kind of hand gestures. Or, if he's on a motorcycle, he may just pull right up beside you and ask that you move onto a side street or exit ramp.

Park your motorcycle, shut it off and KEEP YOUR HANDS WHERE HE CAN SEE THEM! Remember, he has a gun, a uniform and the authority of whatever sovereign state or country you're presently in to back him up. You, on the other hand, have diddley-squat. If he's already had a bad day, the less you do to provoke him, the better.

Do not: Jump off the motorcycle and go for your pockets, backpack, or saddlebags. Remember that for every shooting of a civilian by a cop in this country, about ten or twelve cops get shot by civilians. Don't be stupid.

Do: Display a friendly and cooperative attitude-even if the guy's a dick. Outgun him with karma. Also remember that many jurisdictions now use video cameras on their car stops, so you may be speaking with an officer who's wired for sound and recording everything you say and do. This is for their protection, not necessarily for prosecution, but if you do something stupid, it can come back to haunt you.

Do: Produce your license and registration when he asks for it and if you must go into a saddlebag for it, tell him what you're about to do.

Do not: Lecture the officer about the law, his duties or your constitutional rights—unless you're an attorney, he knows these things as well or better than you do.

If you're female: Do not cry, pout or try to show him your boobs. You'll only insult his intelligence and imply that he's a bully, a beast or a sex fiend, none of which will score you any points.

If you're another law officer, firefighter, or paramedic: By all means let him know that before he starts writing, if possible. Yes, there can be such a thing as professional courtesy, if you weren't going a hundred and smoking crack.

Do: Remember, if you go outside the United States, your constitutional rights have already stopped by the time you reach the border station. You are already a hundred yards or so *inside their country*. Many of the laws on search and seizure that we enjoy do not exist in Mexico or Canada, and in some countries, the cops can seize about anything they want and keep it forever. This includes, but is not limited to, your bike, your personal property and even your person.

Do not: ***Ever*** try to take a firearm or any contraband into Canada or Mexico. At the border station just north of Glacier National Park one year, I had a nice chat with one of the Canadian customs people. He related to me the story of a nice older couple who had come through with their plush motor home the week before. When asked about firearms, they had blithely lied to him. *All* Canadian customs officers are very well trained in body language and recognizing behavior associated with lying and deception. The motor home was searched, perfectly legal in Canada, and two sporting rifles and a Colt Python .357 Magnum were seized. The motor home was impounded and the occupants jailed, to await the arrival of the circuit court magistrate who would judge their case. The customs man smiled slightly when he said the magistrate was "due in a few weeks".

So, you see, getting ticketed is not the end of the world. It usually happens to me when I'm at least 700 miles from home, so I wind up paying it right there, or mailing it in, whatever the pleasure of the local constabulary might be. I always bear in mind that if they caught us every time we went too fast or committed some other minor infraction, none of us would keep a license for long. And since I retired, I'm just like everyone else—cop didn't see me, I didn't do it.

BUYING YOUR BIKE
NEW OR USED?
DEALER OR INDIVIDUAL?
WHAT ABOUT FINANCING?

Just a few of the questions you'll be asking yourself when you get serious about buying a motorcycle. The fact that you bought the book tells me you must be somewhat serious, so let's get started.

New or used? Well, how much money ya got? Or, how good is your credit? Have you ever owned a bike before? What if you buy a brand new shiny one and wreck it because you don't have a clue about what you're doing? We'll cover safety and training later, but for right now you need to ask yourself some serious questions.

New bikes offer some definite advantages: First you get a warranty, which in most cases is extendible. Believe me, that warranty can be a lifesaver. I have had three bikes that required major work, all covered by warranty. Most bikes have one year or 12,000 miles, whichever comes first. Some models that have been in production a long time, such as Kawasaki's Concours have three years, unlimited mileage. These bikes have been in production so long they have been thoroughly de-bugged and are as near "bullet proof" as any machine can ever be.

But, of course, you *pay* for that warranty. Big cycle payments. Interest on your loan. Higher insurance premiums. Higher fees for tags and taxes in many states. We bikers used to use the excuse that we were saving fuel, but now many cars get fuel mileage comparable to or better than most cycles.

If you're just starting out, maybe you should look for a good used bike. Yes, there are such things to be had, but be careful. You must learn to look past the shiny paint and chrome and see what you're really buying. Is the oil black? Is the chain rusty? What color is the brake fluid? If it's brown, it's probably never been changed. What kind of bike is it? How many miles have accumulated? If it has an air-cooled motor and more than forty thousand miles without a top-end overhaul, leave it. You'll be rebuilding it within a year. On the other hand, if it's liquid cooled and has had correct maintenance, it could easily go a hundred thousand miles or more, just like a car. I traded in a 1984 Honda Gold

Wing with 97,000 miles that wasn't even using any oil yet. The salesman bought it and he rode it every day until his untimely death—yes, in a motorcycle accident. That bike had been in all forty-eight of the lower states several times and across Canada twice and had been torn down only once to replace the *stator*, which is the technical name for the outer or stationary windings in the alternator.

Of course, if you buy a used bike from an individual, you don't get the warranty, but for the most part, you don't pay as much for the same make and model as you would to a dealer. The profit motive is usually removed. Often you can save enough to afford a few repairs.

Whenever you're seriously looking at buying a bike from an individual, ask if you can take it to your mechanic and have it checked over. If the guy says no, or seems reluctant, pass it over and look some more. If he says yes and you're really interested in the bike, do exactly that—take it to a good mechanic and have them look it over and do a compression check. A compression check will tell the mechanic (and you) a lot about the condition of wear *on the internal parts of the engine* where it counts. Poor compression means bad valves or worn rings and a top end overhaul coming up real soon. Most bikes can have the "top end" (Rings, valves, honing cylinders) done several times before the bottom end needs attention. The bottom end usually includes bearings for the crankshaft and connecting rods as well as seals and it can be very expensive because the motor must be removed from the frame and the "cases split". "Splitting the cases" refers to the fact that most motorcycle engines have the engine block, or "cases" cast in two halves that mate together around the crankshaft and other lower end parts. Engine removal and replacement alone runs into some serious labor and, as we all know, dealer's labor is pretty steep.

By far, though, the best thing you can do is know ahead of time what kind of bike you want. Do the required reading for that model and know what it's worth. Dealers all have the N.A.D.A. Blue Book that tells them what any bike is worth in their area. It gives average wholesale, average retail and allowances for accessories. They don't deviate from it much and you shouldn't, either. If you've got your eye on a privately-owned bike, call the dealership for that brand and ask someone in their sales department what it might be worth. Most sales people will take the time to help, because they know you'll probably remember their friendliness and bring it to their shop when it needs service or repairs.

Or, better still, get on the Internet and go to the Kelly Bluebook website or N.A.D.A. and look it up for yourself.

Armed with the knowledge of what any given motorcycle might be really worth, you're better prepared to dicker and if it turns out to be more than you wanted to spend, you can shine it on.

Another consideration is insurance. If you buy a new bike and have it financed, you'll have to buy liability, collision and comprehensive, and in some states, medical insurance. You'll probably be paying for a life insurance policy too, so if you kill yourself on it, the loan will pay off. If you can pay cash for a used bike, then you may be able to get by with liability only. You just saved a chunk of money. Chances are if you're in a wreck, you'll be the only one hurt. If your medical insurance from your job covers you, what's the point of buying more medical?

There are many things to consider and

there are some good publications in print to help you and give much more information than I have room for here. Just be aware that there are many good motorcycles that have been garaged and ridden very little. Perhaps the owner died or developed health problems. Many bikes get sold cheaply because of divorces or lay-offs from jobs. It's even possible at times to find a low-mileage bike you can take over payments on that might be half paid for. If I had a choice between paying for half or the whole thing, which do you think I would choose?

Financing your bike can be a challenge, also. Of course, if you buy new, almost all manufacturers offer their own financing. Sometimes you can get better rates than a bank would offer and sometimes you can really get hammered. At least two manufacturers now issue credit cards to their customers to finance their bikes and accessories, riding apparel and whatever else they want. Then they just make a minimum monthly payment. The rate is atrocious, though, like nineteen percent per year. Gee, there's nothing like having twelve thousand dollars or so on your Mastercard. Ouch!

If you buy a used bike, you'll find most

lending institutions will only go about eighty percent on it. You'll need the other twenty percent in cash up front and they will want to see your insurance and there will be a life policy on you until it's paid off.

BIG BIKE OR LITTLE BIKE? *SATISFYING YOUR NEEDS WITHOUT KILLING YOURSELF OR YOUR BUDGET.*

If you've already decided you really *need* that big, pretty Harley Davidson Dynaglide because all your friends have them, fine. If you can afford it, so much the better. I find that Harley riders tend to be a clan unto themselves, rabid defenders of the brand and derisive of anything else as "trash bikes" or "Jap Iron". I find this to be a narrow-minded approach to a sport that should be enjoyable, no matter what brand of bike you might choose.

In my own case, I have no brand loyalty. I really don't think there's a *bad* motorcycle being sold in our country today, government regulations being what they are. All motorcycles sold in the USA must meet certain safety and emissions standards and quality issues tend to take care of themselves. (If you don't believe that, just try and find a Yugo dealership!) Personally, I have had better luck with Hondas than any other brand, but I would not hesitate to buy another brand if the bike strikes my fancy. I give no personal endorsement to anyone. In fact, at this writing, I have just traded in a 2007 Harley Davidson on a 2018 Kawasaki Voyager and it seems to be a very nice ride.

I do most of my own maintenance, (more on that later) and it doesn't take me long to get familiar with whatever I'm riding. What you should be doing is deciding what your own needs are. Are you buying basic transportation, or are you planning road trips? How big are you, physically? Are you particularly short-legged? Are you male or female? Even though I've seen some small women riding some huge bikes, it's usually not their first experience with a motorcycle and it's always a shame to see someone buy more bike than they can handle and become soured on the sport because of too many tip-overs.

Decide what you want from a motorcycle. There are really only a few basic types: Dirt bikes: Tall suspension, knobby tires, rugged construction and not particularly comfortable, but made to handle off-road terrain. Intended for trail riding and motocross racing. Be damned careful when buying one of these used. "Used" often means <u>used up</u>!

Sport bikes: Often called crotch-rockets, these will be the fastest, best-cornering bikes available. They also kill the most riders and therefore have the highest insurance rates. Some of these will run 170 MPH right out of the crate, if you're that crazy. Teenagers tend to *be* that crazy and they think they're immortal. If your teenager wants one of these, get him a body bag along with it and hang on to your hat when you visit the insurance agent's office. The insurance on a teen riding a sport bike is often higher than the payment. This would be a good point at which to put your foot down, Mom and Dad.

Cruisers: Many of the Harleys and a lot of the Japanese bikes fit into this category. Usually V-twin powered with a semi-chopper look, they rumble up and down the avenues at night, their owners profiling and showing off machines, chicks and leather. More chrome than you could polish in a week and in many cases not much in the way of comfort, although that can be cured with the addition of a few accessories. Prices can run from eight to thirty thousand dollars. How much do ya wanna spend?

Tour bikes: Again lots of Harleys and

some Japanese bikes. Big bikes, with windshields, fairings, saddlebags and trunks. Bikes you can actually travel on and not just across town. Many couples go for weeks at a time and camp or do the motel thing. Most of these go for eleven to twenty-five thousand new. When you look for used bikes, you'll find a lot of these, usually over-decorated with about a ton of chrome and lights, selling for fairly reasonable prices. If you don't mind all the decorations, or you don't mind stripping them off, you might find a good buy. But, be aware: Often a bike that has been parked for a long time will offer a whole set of unique problems: Seals dry up, grease sets up into glue, wiring gets chewed by mice and gasoline turns to greenish glop…just be careful.

Dual-purpose: Most of the manufacturers offer at least one of these. A bike that's equally at home on the highway or the back trails. At least that's what the brochures say. They're made for the guy who can't decide what he really wants out of a motorcycle. Or the guy who really needs two bikes but can only afford one. Or maybe for the guy who just gets a wild hair whenever he sees an interesting gravel road. A little "spendy" for what you get, usually pretty powerful and some even have dual-range transmissions. I have a friend who takes motorcycle tours into Mexico. For the roads down there, a dual-purpose bike such as the Kawasaki KLR is a good choice.

So, what's it gonna be, Mac? Big or small? Cruiser or crotch rocket? You need to take a realistic look at what you'll be using the bike for, what the weather is like where you live, (can you ride year-round or only three months or so? Do you need a fairing and bags?) and, whether you'll be taking a wife or girlfriend along. If you buy a bike that's too small, I guarantee you'll trade up within a year and probably lose money, but then, that's part of learning, too. I did it that way, starting with a 90cc Bridgestone street bike that I bought for $145.00 from a guy who bought out a defunct dealership in Newton, Kansas. I carried it home in the trunk of my wife's Olds convertible. In a few months, I was up to a 305cc Honda and then, within a year I was riding a 74 cubic inch Harley Duo Glide. It doesn't take long to move up, believe me.

Whatever bike you choose, you'll need some training and you'll need to know a lot of things you don't know if you've only driven cars. We'll talk about those things in the next chapter.

RIDING
THE TWELVE SECOND RULE, THE TWO SECOND RULE AND DYING WITH YOUR BOOTS ON.

Probably the worst assumption people can make is that they should be able to ride a motorcycle based solely on what they already know. It goes something like this: **I'm a good, accident-free driver. I can drive a car and I can ride a bicycle, so I should be able to ride a motorcycle, right?**

Wrong! Actually, nothing could be farther from the truth. Motorcycle operation requires very special skills and one should not just jump on and go out and play in the traffic any more than one should go crawl into a helicopter and try to fly without instruction. Even if you *did* stay at a Holiday Inn Express last night.

That having been said, I'll confess that the aforementioned procedure is *exactly* how *I* did it, back in the late 60's and I got away with it for several reasons: First, I was not a teenager—I got all the wild driving and stunts out of my system with muscle cars before I ever bought my first motorcycle. Second, the traffic then was nothing like we have now. Third, I really wasn't all that good with a motorcycle until I got formal training from the police department some thirteen years after I started riding. Oh, yeah, I *thought I was good* and I was pretty much accident free, but I didn't really handle a bike all that well—and I was lucky.

A good source of information on motorcycle accidents and their causes is the Hurt report, published by Dr. Harry Hurt of USC, Berkeley. In a comprehensive study of 800 motorcycle accidents, his team found that a disproportionate number of motorcycle accidents involved first-time riders and riders who were self-taught or taught by a relative. We'll talk more about training in a later chapter, but suffice it to say, you do need it and the more the better.

Right now, though, let's talk about what it takes to survive, and remember, I speak from experience—I've been doing it for fifty years.

First, know your mount. Inspect it before you ride—not just once in a while, but every time you get ready to take it out, even if it's just up to the corner store. That nail you find in your tire might have been the blowout that killed you. We'll get into inspecting the machine by the "T-Clock" method in the safety chapter.

In the heading of this chapter I mentioned the two-second rule and the twelve-second rule. You may have heard of the first. It's the rule that says you stay two seconds behind the vehicle in front of you as a safe following distance. But bikes can't stop as fast as cars, right? **Wrong again!** *Bikes will stop faster, if you use the brakes the way they were designed to be used.* Most people don't, though, and we'll be covering that, too.

The twelve-second rule goes like this: As a motorcycle rider you should be observing everything for a distance of twelve seconds down the road. Obviously, at twenty MPH, twelve seconds isn't very far, but at a hundred, it's a long ways! Not that I'd ever ride that fast. Not me. (Hah!)

Why twelve seconds? Take the time it takes to see and recognize danger, add in your reaction time, then the time it takes to decide what action is appropriate. Do I brake or swerve, or both? Add in the time it takes to execute the maneuver safely and you'll see that your twelve seconds are about up.

In the Wichita Police motorcycle school, they taught us that if you're going seventy MPH and something steps into your path within a hundred yards, you will hit it. They also said that the human body cannot survive striking an immovable object at forty MPH. The first is true because your reaction time will take too long for you to decide the appropriate action and still have time to execute it. The second is true because the G-forces associated with a dead-stop impact from forty MPH will burst your internal organs and you will die.

Learning to properly brake and swerve the motorcycle are the two factors, along with good judgment, that will be most likely to save your life. Using good safety equipment is another factor that can make a big difference in whether you live to ride another day. And the final factor is attitude. Like so many other things in life, attitude can mean life or death. The *professional* motorcyclist adopts the attitude that everyone else is drunk, crazy or just out to get him. He has infinite patience with the mistakes of others and he knows all the cute ways "cage" drivers will try to kill him. He takes all this in stride and still enjoys the ride. There is no room for bad attitudes or "road rage" when you straddle a bike. Not if you want to live. Forget Hollywood and their "bad biker" flicks. I always have to laugh when I see one of those movies where the bikers are terrorizing the town, even though such tripe gives us all a bad reputation.

Think about it for a minute; which is a better weapon, a car or a chain? Have you ever tried to swing a chain heavy enough to bust out a windshield, while trying to control a motorcycle? If a guy really tried this movie crap, Ol' Granny in her Taurus would probably stick a fender up his butt and drive on, leaving him in a smoking heap in the ditch.

The truth is that you have to protect *yourself.* You do this by picking where you will ride and, basically, covering your ass. People tell me all the time, "Well, ya can't drive for the other guy." That, too, is a common misconception. The fact is, you *can* drive for the other guy, and you *must,* in order to survive. I've been driving for the other guy for about fifty years.

I like to ride in the left half of the right-hand lane, or the right half of the left-hand lane. That way, a lane change only involves moving the machine about six to eight feet. If

I have cars around me, I try to pick my spot so as to give myself as much distance as I can from them. The idea is to build a "circle of safety" around yourself. The faster you go, the more the circle elongates into an oval shape. Whenever another vehicle starts "bumping" that invisible circle, in other words, getting too close for comfort, don't wait, move! Find some way to increase the size of the circle. You might have to speed up, slow down, or change lanes, but don't wait to see what happens next. You might not like the result.

I tend to watch the front wheels on cars as I pass them. If the wheel starts to turn, I know the driver hasn't seen me and he's possibly going to change lanes and take me out.

I always avoid the "grease strip" in the center of the driving lane, especially near intersections. This is the area where the car engines drip the most, as they set idling and waiting on the light. This can be very slick and hazardous to a motorcyclist, especially when you get a little rain on top of it.

I was trained to ride *through* the flow of traffic rather than *with* the flow. This may make you more vulnerable to speeding tickets, but it means you spend less time in anyone's "blind spot". (There really is no such thing as a blind spot. People are just lazy and they don't turn their heads or use their mirrors. When you're texting on a cell phone, the whole *world* is a blind spot.) I try to watch other drivers and look for the eye contact that tells me they've seen me. If I don't see the eye contact, I'm going for my brakes. With the proliferation of cell phone idiots out there now, I find myself going for the brakes a *lot*.

I tend to brake harder with my front wheel and more gently with the rear. Seventy percent of a motorcycle's braking force comes from the front wheel brake. On a modern-day motorcycle, there are usually two brake rotors on the front wheel, making the front brake very efficient and fade-resistant. **The only way it will hurt you is if you don't use it properly.** You'll hear old-time riders tell you, "Stay off thet there front brake! It'll throw ya plumb over the handlebars." Yeah, right. Maybe if you're riding an English bicycle. Not on a motorcycle. The only hazards of using a front brake aggressively are braking on treacherous surfaces, such as sand or that oily strip you shouldn't be riding on anyway, and "snatching" the front brake and getting a front wheel lockup before the weight of the motorcycle transfers to the front wheel. In that case, the front fork can snap to one side or the other, throwing the rider off.

In an emergency stop, the drill goes this way: Keep your head up, eyes on the horizon, front brake applied harder and harder until the stop is reached, rear brake "feathered" just enough to keep the back end from trying to pass the front. The idea is to scrub off as much speed as possible. Even if you crash, the slower you're going when it happens, the less injury you're going to sustain.

In the Police school they taught us that *every* time you take a spill you will sustain *some injury*, even if it's only a little bump, bruise, or "road rash". Don't make the mistake of thinking that road rash isn't a significant injury. There are cases on record of people *dying* from road rash. It's an abrasion that scrapes into the layers of your skin, which is your body's first defense against infection. Like a burn, road rash is easily infected and if you get a Staph or Strep infection in a good-sized patch of road rash it can be deadly.

The best cure for road rash is, of course, prevention. The rule is this: **Wear as much**

protective clothing as you can stand. We'll be covering this in the chapter on riding apparel.

Perhaps the one thing that most quickly points out the amateur motorcyclist in traffic is the way a rider starts and stops the motorcycle. I see them all the time, walking the bike the last fifteen feet up to a stop sign, paddling their feet like a baby duck when they start off, or riding a half block with their feet dangling down. I always ask myself, "If they wanted to walk, why did they buy a motorcycle?"

The professionally trained rider knows that dangling, paddling feet are an invitation to foot and ankle injuries. He learns to stop the machine, using *both* brakes, on a pre-selected spot, with only his left foot down. The right foot remains on the rear brake pedal. When he starts off, he raises the left foot immediately, bringing it *forward* and up onto the peg or floorboard. A good exercise is to lay an old car tire (one that's not mounted on a wheel) flat on the pavement in a parking lot and practice stopping next to it, placing your foot inside the tire as you stop. Then start off, lifting your foot out of the tire as you begin to move.

As a confidence builder and to demonstrate to yourself how long it actually takes for a motorcycle to fall over, try this: Stop and put both feet down. Now, raise both feet at the same time onto the pegs and just drive off. Don't let the machine move until after your feet are on the pegs. It sounds risky, but actually it's very easy to do and it looks pretty cool to those who don't ride. They'll wonder if you've got a gyroscope hidden in your shorts!

Another good thing to practice is "slow riding". This is so difficult for many riders that it's usually a main event in the games at motorcycle rallies. It's really not that difficult to ride super-slow, once you get the hang of it. First, keep your head up and your eyes on the horizon. Learn to *feel* the machine and anticipate which way it's going to try to tip. Using just brief touches of the clutch and throttle, keep the machine rolling, but just barely. Whenever it starts moving too fast, ride the rear brake a little. This is one time when you try to avoid the front brake. With a little practice you can keep the machine rolling and stable while going slower than normal walking speed. This comes in really handy in those traffic jams where everybody keeps moving up a few feet at a time for half a mile. You'll be the one biker who doesn't *walk* that half-mile.

The true gauge of a skilled rider is not how well he handles a bike at high speed. Anyone can go fast. But watch when the going is really slow and you'll see who has accumulated skill and who is the rank amateur.

Sometimes, when I'm around a group of riders, I'll say something about "counter-steering", just to see how many blank looks I'll get. What the heck's that? The fact is, when the handlebar on a motorcycle traveling at any appreciable speed is turned to the right, the machine will go left and vice-versa. There is a whole set of physical laws that show why this happens, but I always tell students that it's just "magic".

If you've ridden for some time, you have unconsciously been counter-steering and you probably didn't even know it. Once you realize how this works, it becomes a definite advantage when you have to swerve suddenly. A sharp push on the *correct* side of the handlebar will quickly swerve the bike safely with a minimum of fuss and allow you to miss that muffler lying in the road or your neighbor's cat.

It's as simple as it could possibly be if you just develop the mindset: **Press left-go left, press right-go right.** And of course a press in the opposite direction recovers the bike from the swerve and back onto the straight track.

Swerving and braking. If you don't learn anything else, you need these skills. In the chapter on training, I'll tell you where to get them.

GETTING TRAINED
THE BEST AND WORST WAYS TO LEARN.

By far the worst way to learn to ride is the way I did it back in the sixties: I bought that old Bridgestone, got on it and went out to play in the traffic. Not the recommended procedure nowadays. Remember, the traffic wasn't anything like it is now. And I had ridden other people's machines before I bought one. Still, I was lucky and I was really careful.

By the time I got any formal training, I had been riding my own bikes for about thirteen or fourteen years. In that time, I'd had only one spill and I'd never collided with any other vehicle. But it was still amazing how many bad habits I had to "unlearn" and how little actual skill I had accumulated in all my riding: I was street-wise, but I didn't really handle a motorcycle well.

The second-worst way to learn is from a relative. They're too personally involved and their patience will wear thin if they don't feel you're learning "fast enough". When I taught for the Motorcycle Safety Foundation, I quickly found out that everyone learns at a different rate and that the level of concentration and length of attention span vary greatly from one individual to another. A person who is used to teaching does much better than a relative, even if he only has one student. In the case of motorcycle training, the student is being asked to absorb a great deal of information in a short amount of time, and apply it immediately to his riding. Very difficult for the student and exasperating if the teacher is a husband or boyfriend or your dad. Personal issues get involved and more often than not, you wind up with a disaster for all involved. A relative also is going to freak out every time you tip over or fall off. A professional instructor realizes that students are going to take a spill once in a while. That's why the training is done at low speeds in parking lots and never out on the streets. That's also why professionals will not train anyone unless they wear <u>all</u> of the required safety gear.

If I had it to do nowadays, I'd go to my dealer and ask about finding a training course. Most dealers support the Motorcycle Safety Foundation (MSF) by providing motorcycles, maintenance, parts, fuel, or whatever they can afford. Again, it's just good business. The more riders there are, the more bikes the dealer gets to sell and service.

The MSF has two courses they regularly teach. The standard course takes two evenings and a weekend, beginning on a Thursday afternoon and ending late Sunday. The motorcycles are provided along with all study materials. The student provides the cost of the course, and his protective gear. They require a helmet, eye protection, full-fingered gloves, boots or shoes that cover the ankle, long sleeved shirt or jacket and jeans or pants that cover the entire leg.

The advanced, or Experienced Rider Course is an eight-hour refresher that's done in one day, usually a Saturday or Sunday. I know riders who take this course every spring, just to stay sharp at their skills. Personally, I never put my bike away for the winter, preferring to bundle up and ride if the streets are clear and dry. In Kansas, I have about thirty days in the average year when it's impossible to ride. The rest of the time, I'm on it. For those riders who store their bikes over winter, we'll have some tips on storage in the maintenance chapter.

MSF instructors work a full week of long days to get their ratings and most take the program very seriously. If you are unable to find an instructor near you, you can write:

Motorcycle Safety Foundation:

2 Jenner Street, Suite 150

Irvine, California, 92718

Or, find them on the Internet.

WHAT ABOUT MAINTENANCE?
THE MANUAL, THE ATTITUDE AND SOME DO'S AND DON'TS.

You say you're a do-it-yourselfer? Well, so am I. It just *kills* me to pay someone else to do something I can do just as well, or maybe even better. I'm one of those guys they call anal-retentive. I tend to be picky as hell about things like maintenance.

Over the years, I've learned to do wiring, sheetrock, linoleum, roofing, ceramic tile and dozens of other useful things, including my own motorcycle service work.

If this is the way you are, too, may I suggest the following: First, get the manual. Not the owner's booklet they give you with the bike. The big one. The one you have to order. The same one the dealer's mechanic uses. It'll have a part number and you can order it right over the parts counter. Expect to pay upwards of sixty bucks or more. That's okay, it will save you that much in the first year, easily. Or, if you get lucky like I did, you might find a service manual in PDF format on the Internet that you can download for nothing. Then, you can print out whatever pages you need for reference in the shop.

Next, make sure you have a good set of tools. Don't try to "make do". This will only cause a lot of skinned knuckles and rounded-off nuts and bolts. Of course, you can't afford every special tool the dealer has and most dealers won't loan tools. But if you're looking at fixing a problem that requires special tools, you may be too far into it, anyway. The trick is to know when to attack the work and when to back off and give it to a certified, factory-trained technician.

The average guy with a good set of tools should be able to accomplish his own routine oil and filter changes, chain adjustments, control cable adjustments and that sort of thing. Lots of guys even take their own wheels on and off and change tires. I take my bike in to the dealer for tire work, because they have a tire machine and can do it in a tenth the time I can. And whether I take in the bike or just the wheel, they charge the same.

When it comes to tearing down engines and complicated assemblies, I go with the dealer or professional shop every time, and here's why: Once they touch a wrench to my machine and work on the engine, *they have to stand behind whatever repairs they make.* If it's not right, it goes back as many times as it takes until it is. And often, the factory-trained tech knows things that aren't necessarily in the manual.

I worked for a time as a service manager for a dealer in Wichita and I saw a lot of engines come in being carried in crates and baskets. The owners had torn them down, then found they didn't have the tools, gauges and skills to rebuild them with any hope of success. I once had a customer bring in a Honda V45 Magna engine that looked like a grenade had gone off inside it. He had "overhauled" it himself and probably did a passable job, except he didn't know that the connecting rod bolts are not reusable. When they are properly torqued, they "stretch" slightly, and once disassembled, the engine requires placement of new bolts in this one critical area, every time.

The customer's engine failed at high speed and the broken pieces flailed around inside, beating hell out of everything and making the next rebuild *very* expensive. He was lucky he wasn't killed. Had the engine decided to totally seize, it could have locked the rear wheel and caused a fatal crash.

So, when it comes to maintenance, as with your riding, know your limitations and don't exceed them.

Many riders prefer to store their machines in winter, but few really know how tough this can be on a bike. Batteries slowly discharge, gasoline sours and congeals into glop and exposed metal begins to rust. What can a guy do to insure that when spring rolls around the bike will be ready to ride, rather than trailer to the shop?

First, let's consider the battery: A storage system for electrical energy that depends on a chemical reaction of sulfuric acid on lead plates to hold a charge. If used regularly and kept "wet" by topping off with distilled water, it can last a long time. Neglected or poorly serviced, it may not last even one year. Make it a habit to check the water level in the battery every week. I don't care if it has some kind of rinky-dink indicator that's supposed to show when it gets low, or that it may be hard to get to. Batteries are expensive. Nowadays, many bikes come with sealed batteries, which make battery maintenance a thing of the past. Just like car batteries, they last however they last and there's really nothing you can do about it, but replace the battery whenever it's time.

Hint: Many cycle batteries can be found at discount houses such as Wal Mart and Sam's. The dealer will tell you *his* battery is better. Don't believe him. They all come from the same factory in Japan. The labels are added later. It's the same battery for about half the money and you *can* replace it yourself. Just remember, unlike a car, most bikes don't have a strong enough alternator to "bring up" a weak or dead battery. Charge it fully at a slow rate before you put it in service, or it will die an early death.

How do you know when a battery is fully charged? That's easy, too. While you're at Wal-Mart, go to the automotive section and buy a little gizmo called a hydrometer. It has a squeeze-bulb on one end and a glass or plastic tube you insert down into each cell of the battery. It also has three or four little colored balls inside the tube. When you draw some of the electrolyte (acid) up into the tube, the balls will float. The higher the *specific gravity* of the electrolyte, the more little balls will float. This is a good way to find a dead cell in a battery, too. The higher the charge in the battery, the higher the specific gravity of the solution. So, if all your balls float, you're really charged up. (Did I really say that?)

Next, buy a **battery maintainer**. I got mine for twenty-nine bucks at Sears and wired the leads permanently to the battery. Whenever I know the bike's going to sit for more than a few days, I plug it in. The maintainer senses when the battery begins to discharge and "trickle" charges it automatically. So when I'm ready to ride, the bike's ready, too. In the winter, the battery maintainer is a must.

Next, look to your carburetors. They have float bowls with gas in them. If it's not drained and it sits for several months (or years) it will evaporate, leaving a substance that resembles green jell-o. Then you start the bike and suck all that crap into your jets and, bingo! You've now got a crappy-running motorcycle and an expensive carb cleaning job to pay for.

What are jets? Sounds like a high-performance aircraft, I know. Here are some carburetor basics. All carbs have jets, or orifices that fuel is sucked or forced through to create a mist when mixed with air. This is a very explosive mixture and this is where your engine gets its power. This mixture is compressed in the cylinders, ignited by a spark plug and the result is all that cool noise and power. But, one set of carb jets can't handle all ranges of power we demand from a modern engine. So most carbs have about four "circuits" or sets of different-sized jets, a set for idle, a low-speed, a mid-range and a high-speed. Carbs also have either a "butterfly" air valve in the top, or a slide valve with a tapered "needle" that seats into the main jet to control air flow.

Carburetors on modern motorcycles are set at the factory for optimum performance and fuel economy and to meet EPA air-quality emission standards. Almost all manufacturers have found it necessary to build a California model and a 49-state model of their machines, because California has the most stringent air-quality standards of any state. Check out the smog in Los Angeles during rush hour and you'll know why.

New motorcycles tend to be pretty quiet, too, because of noise pollution standards. Of course, the first thing you'll want to do is make the bike "sound better". You'll want to put on a set of after-market pipes, which will screw up all those factory settings we just talked about. A less restrictive exhaust means more airflow through that engine and the fuel metering in the carbs is still the same. This creates a lean mixture, which will cause low power at certain ranges and a hotter running engine. The solution? Get the carbs re-jetted. Yeah. More money. This is something you don't want to try and do yourself, unless you really know what you're doing. After re-jetting, when everything sounds "pretty", or maybe just obnoxiously loud, and everything's running as good as it was before you messed with it, you'll find your fuel economy just went to hell. I once "piped" and re-jetted a '97 Honda Shadow and went from 42 miles per gallon to 32 miles per gallon! Sure sounded sweet, though.

Okay, I got off track a bit, but when you want to store the bike, take care of those carbs. Here's what to do: On each float bowl, on the bottom of the carb, there's a drain screw. There's also a tube running off the float bowl, down through the frame and pointing to the ground. Turn off the gas valve at the tank and put a towel on the garage floor to catch the gas. (If your bike has a fuel pump, there may not be a fuel valve.) Then, one by one, back out each of the drain screws until each float bowl is drained, then close them. Then go to the auto parts store and buy a small can of gasoline stabilizer. Follow the instructions on the can. Remember, your tank probably only holds five or six gallons. I would have the tank just about full, then add the appropriate amount of stabilizer, per the instructions on the can. When spring comes and the bike comes out of storage, turn on the gas valve and let the carbs have stabilized gas and start her up. Then I'd try to go through that tank of gas as quickly as I could, just to get fresh gas into the fuel system.

Of course, if your bike is fuel-injected, all this carb information becomes out of date and therefore, useless.

Lastly, I like to cover the bike, just to keep dust and grime off. Trouble is, some cycle covers may trap moisture underneath, causing chrome to pit and rust. You can get moisture absorbent compounds at most hardware stores. They will pull moisture out of the air and are usually intended to take dampness out of basements. A can of this stuff sitting on the floor under the cover will help keep the bike dry. Check it once a week and replace it when it becomes soaked with water. Some bike manuals even recommend coating all metal parts with oil when storing for long periods of time. This is messy but effective. When new motorcycles are shipped from the factory, they're sprayed with a preservative that's mostly a light grease. It's a mess to get off the bike during dealer prep, but the manufacturer never knows how long that bike may sit in a dealer's warehouse before it's set up and sold. I don't coat my bikes with anything because I never store them for any length of time.

A bike's worst enemy, though, is the sun. Through our use of hydrocarbon fuels, we have thinned the ozone layer to the point that ultraviolet light will ruin paint, rubber and plastic much more quickly than most folks realize. I keep my bike garaged at home and when I was still working I was fortunate enough to have a parking garage at work, too. When I go on trips, I take along a good quality nylon cover. When I park it, I give it time for the pipes to cool so they won't melt the cover material, then I clean the windshield and lights and get it covered up. The bike can look good for years with a little care. I never use waxes with any abrasive content. Repeated use of these will thin the paint, eventually causing clear coat and striping to disappear. I also never use chrome polish. I wash the bike using cool water and a little Simple Green on the greasy parts. I use a natural chamois skin to dry it and then buff it with a dry towel and Maguire's Instant Detailer.

The last thing to remember is that if you bought a new bike, there are certain dealer checks that are required to keep your warranty in force. Don't skimp on these! Have them done on time and keep the receipts. The phrase you never want to hear from a dealer is, "Sorry, you screwed up and voiced your warranty."

RIDING APPAREL
THE RAINSUIT, THE COLD WEATHER RIDER–HARDCORE VS FAIR-WEATHER RIDERS.

What's the best material for a biker to wear? In spite of all of our technology and breakthroughs in producing man-made fibers, I will still go with leather and here's the reason why. All man-made fibers, from Nylon to Kevlar and everything in between, will melt from heat. When you slide on pavement, friction causes heat and the material is scraped and *melted* away. Leather, on the other hand, will scrape off, but not melt. So I still go with what God chose to wrap his cows in. Nothing like good ol' cowhide. I have chaps, pants, jackets and the whole mess and I wear them whenever it's not too hot.

I <u>never</u> ride in shorts, period. If you have pretty legs and you want to show them off, fine, but be aware that when you fall, (and eventually you *will* fall) you'll scar yourself worse if you happen to be in shorts. I don't wear them for another reason: My skinny, hairy white legs tend to scare the cattle when I ride in the country, and they stop giving milk. The County Extension Agent and a representative of an animal rights group contacted me about this. They were pleasant enough, but they assured me that if I did not cease and desist, a court order would be forthcoming. You believe that, right?

I always wear boots and gloves, even in summer. Hand and foot injuries are more painful than any other kind because of the concentration of nerve endings in the hands and feet. When you take a spill, there is a natural tendency to put your hands out to break your fall. Every time I have ever fallen from a motorcycle I have ruined a pair of gloves and I've always been glad afterwards that I had them on. Sure, they're hot in summer and your hands get sweaty, but if I couldn't stand a little discomfort to protect myself, I wouldn't ride a bike.

I also never ride without some form of eye protection. My bikes always have windshields, but that really isn't enough. You'd be surprised at the amount of bugs and other crap that gets around a windshield. I've been nailed right in the face by rocks, June bugs, Bullet bugs, Mashed Potato bugs and those ones that look like somebody threw a spoonful of mustard in your face. There's also an Epoxy bug that's extremely hard to get off windshield plastic. A couple of weeks ago, on the Great Plains of Kansas, while riding in moderate rain, I got nailed right square on the nose by a hailstone! Dude, that hurt!

If you're going 60 MPH and meet a truck going 60 MPH the other way and he throws a rock, it's interesting, to say the least. The rock will be going the equivalent of 120 MPH or better when it contacts your body, and the first time it happens, you'll swear you've been shot. Your first impulse is to pull over and have a good cry, but of course we bikers are too macho for that. Usually I find a good long stream of cursing to be therapeutic.

People often ask, "But what do you do when it RAINS?" Like rain was the end of the world. Hey, c'mon, it's just rain…I answer, "I pull over and suit up and go on down the road." This is true to a certain extent. I don't "do" thunderstorms. I don't like lightning and I have seen enough tornadoes in Kansas to know better than to ride into violent weather. But if it's just a little rain, and you're a thousand miles from home, what option do you have? If you live somewhere like Seattle, you'll want a really good rain suit. Out here in the wilds of Kansas I get caught in rain maybe four times a year. I buy my rain suits at Wal Mart for about nine bucks and almost always manage to melt a hole in the right leg on my exhaust pipe. It gets replaced at least once a year.

The real secret to long-distance rain riding, though, is keeping your feet and hands dry, along with the rest of you. It doesn't matter if your boots get soaked all the way through as long as you keep your *skin* dry. Once the skin becomes wet, your body heat is quickly lost and you'll be miserable. If you don't want to mess with rubber boots, do this: Find some plastic bags just large enough to fit over your feet. Pull them on over your socks and put your boots back on. Your feeties stay dry and comfy and your disposition remains tranquil.

For the hands, go find a service station supply or a place that handles specialty rubber items. After you get finished admiring the blow-up love dolls and 'Texas latex' giant condoms, ask them for chemical handler's gloves. Get them large enough to go over your regular gloves. The cuff will reach halfway up your arm and they'll be a little clumsy until you get used to them, but they'll keep the hands dry.

Then there's the face. I carry a "rip-off" face shield, still in its protective foam sleeve, stashed in the bottom of my saddlebag. These just bend to the curvature of the helmet and snap in place. They give another layer of protection and when they're new and shiny, the rain beads up and flies right off.

What happens if you're out on a ride and it suddenly turns cold? Temperature drops twenty-five or thirty degrees and all you have is a light jacket? Stop and buy a newspaper. No, don't sit there and read it until the weather warms up. Use it to line your clothing. Slide sheets of newspaper down into your sleeves, or buy some duct tape and go in the restroom and tape the newsprint around your arms, legs and torso, then *put your clothes back on*. Did I have to tell you that last part? Well, I dunno…anyway, you'll be surprised how much warmer you'll be and nobody ever needs to know what that crackling sound was when you walked back past the checkout.

What did I mean when I spoke about "hardcore vs fair-weather" riders? Just this: I know guys who spend upwards of four hundred bucks a month on a cycle payment, at least four hundred a year on insurance, plus tags and taxes, just to go ride around town on Saturday nights and play biker. I call these types biker wanna-be's. If there's a cloud in the sky they take the car and they consider a ride to the lake, 25 miles away, a "road trip". If that's your bag and you can afford it, power to ya. I can't.

My bike payment runs one seventy-three a month and I average around ten thousand miles a year on my machine. I get a lot more "bang for my buck" and I enjoy it when people think I'm crazy. So I guess it's whatever blows your skirt up, huh?

WHAT ABOUT SAFETY?
LOOKING GOOD, BEING SAFE AND NOT FEELING RIDICULOUS.

Personally, I think you can carry safety to an extreme. I figure, if I'm riding in a manner which allows me my "circle of safety", my lights are on, I'm positioned correctly and I'm paying attention, if someone's going to run over me, one of those stupid orange day-glo vests won't save my ass. Chances are, if the cage driver's that blind, he'll just figure I'm a traffic cone or one of those orange plastic barrels and cream me, anyway.

However, I do see guys who seem to think it's extremely macho to fix their lights so they can turn them off in the daytime. Again, that attitude. They rig the lights, turn them off, then get upset when somebody doesn't see them. Gives them the opportunity to get crosswise with other drivers, I guess. Then they can yell and cuss and go to the bar and tell their buddies about the old granny that tried to run them off the road. I put these guys in the same class with the Neanderthals who take their turn signals off and make their old lady signal for them with her hands.

Now think: Do driver education instructors still *teach* hand signaling? I have my doubts. If they do, it's a five-minute dissertation, quickly forgotten by the student, who has about a million other things to deal with. I notice they no longer teach kids to drive a standard shift, or to parallel park, so why would they teach hand signals?

The other thing to remember is that as soon as most cage drivers get their license, they stop learning right there, as far as good driving habits go. They'll continue to learn the bad ones, though.

So, if you don't flash a turn signal in their face, they have no clue as to what you're trying to tell them. You'll see motor cops using turn signals *in parking lots*, out of habit and because they know how many drivers there are out there who are legally blind, mentally incompetent, stressed beyond measure, or just abysmally stupid and there's no point in making one of these jerks mad. Remember, some of them even carry guns.

Earlier, I mentioned the T-Clock inspection. When you do this procedure, you'll make sure your lights are all working. Here's how it goes, and by the way, this is a straight rip-off from the Motorcycle Safety

Foundation's manual:

T-CLOCK INSPECTION

T= TIRES & WHEELS

Tires: **Condition—**Tread depth, wear, weathering, evenly seated, bulges, imbedded objects.

Air Pressure— Check when cold, adjust to load/ speed.

Wheels: **Spokes—**Bent, broken, missing, tension, check at top of wheel. "Ring" OK—"Thud", loose spoke.

Cast—Cracks, dents.

Rims—Out of round/true, + or – 5mm. Spin wheel & index against stationary pointer.

Bearings—Grab top and bottom of tire and flex: No freeplay (click) between hub and axle, no growl when spinning.

Seals—Cracked, cut or torn, excessive grease on outside, reddish-brown around outside.

C=CONTROLS

Levers: **Condition—**Broken, bent, cracked, mounts tight, ball ends on handlebar levers.

Pivots—Lubricated properly.

Cables: **Condition—**Fraying, kinks, lubrication: ends and length.

Routing—No interference or pulling at steering head, suspension, no sharp angles, wire looms in place.

Hoses: **Condition—**Cuts, cracks, leaks, bulges, chafing, deterioration.

Throttle: **Operation—**Moves freely, snaps closed, no revving.

L=LIGHTS

Battery: **Condition—**Terminals clean and tight, electrolyte level, cables held down securely.

Vent tube—Not kinked, routed properly, not plugged.

Lenses: **Condition—**Cracked, broken, securely mounted, excessive condensation.

Reflectors: **Condition—**Cracked, broken, securely

mounted.

Wiring: **Condition**—Fraying, chafing, insulation.

Routing—Pinched, no interference or pulling at steering head or suspension, wire looms and ties in place, connectors tight and clean.

Headlamp: **Condition**—Cracks, reflector, mounting and adjustment system.

Aim—Height and right/left.

O=OIL

Levels: **Engine oil**—Check warm on center stand, dipstick, sight glass.

Hypoid Gear Oil— Transmission, rear drive, shaft.

Hydraulic Fluid— Brakes, clutch, reservoir or sight glass.

Coolant— Reservoir and/or coolant recovery tank, check cool only!

Fuel— Tank or gauge.

Leaks: **Engine Oil**— Gaskets, housings, seals.

Hypoid Gear Oil—Gaskets, seals,

breathers.

Hydraulic Fluid— Hoses, Master cylinders, Calipers.

Coolant— Radiator, hoses, tanks, fittings, pipes.

Fuel— Lines, fuel taps, carbs.

C=CHASSIS

Frame: **Condition—**Cracks at gussets, accessory mounts, look for paint lifting.

Steering Head bearings—No detent or tight spots through full travel, raise front wheel, check for play by pushing/pulling forks.

Swing Arm Bushings/Bearings— Raise rear wheel, check for play by pushing/pulling swing arm.

Suspension forks—Smooth travel, equal air pressure/damping anti-dive settings.

Shock(s)—Smooth travel, equal pre-load/air pressure/damping settings, linkage moves freely and is lubricated.

Chain or belt: **Tension—**Check at tightest point.

Lubrication—Side plates when hot.

Note: Do not lubricate belts.

Sprockets— Teeth not hooked, securely mounted.

Fasteners/Threaded—Tight, missing bolts, nuts.

Clips— Broken, missing.

Cotter Pins— Broken, missing.

K=KICKSTAND

Center stand: **Condition—** Cracks, bent.

Retention— Springs in place, tension to hold position.

Side stand: **Condition—** Cracks, bent, (safety cut-out switch or pad, if equipped.)

Retention— Springs in place, tension to hold position.

Do I do this whole rigmarole every time I ride? Hell, no. You won't either, but after a while, you'll become familiar enough with your machine, you won't have to. It would be a good habit to get into for a while, though. You'll form your own bad habits soon enough, without any help from me.

CHOOSING A HELMET
CHEAP OR EXPENSIVE–THE TEN-DOLLAR HEAD.

"If ya got a ten dollar head, go ahead and buy a ten dollar helmet." I can hear Jim Stansberry telling the "motor babies" that, even today. Jim was a Lieutenant on the Wichita Police Department and, along with Charlie Lauffer, one of my instructors at motorcycle school. I credit their training with saving my life dozens of times.

Jim's feeling was, if you want to ride without a helmet, you take your chances and you live with the risk, but if you opt for the protection of the helmet, why buy a cheap piece of crap?

I've never bothered with a helmet that cost less than a hundred bucks. I believe in hand-laid fiberglass and serious retention features. If it comes off on impact, it won't do a thing for you.

For a while, there were a lot of polycarbonate helmets out there, but they seem to have mostly disappeared now. They were cheap and the material tended to break down from ultraviolet light. Eventually they discolored and became brittle, usually long after they should have been discarded, anyway.

The rule with helmets is: Never use a helmet more than two years. The internal components will have broken down from heat, sweat, hair oil, or whatever and even if the shell has good integrity, that's only one of the protection factors.

Helmets save your head by lessening impact to your skull. They do this by *destruction of the material of the helmet.* This cuts down those G-forces we talked about earlier and increases your survivability factor. On impact, the fiberglass shell crushes, then the inner foam layer. If the components have deteriorated, they will not do an adequate job of energy absorption and guess where that energy goes? Right into your noggin. Excessive G-forces on the skull will cause the brain to shift and slam around inside its own protective *dura*, which is the layer between the skull and the brain itself. This can result in concussion, skull fracture and immediate or eventual death.

Folks who don't want to wear helmets come up with some pretty lame excuses: "When I have it on, I can't hear." *I find I hear just as well, because I eliminate the wind noise over my ears. I know guys who won't wear a helmet, but will wear earplugs. Go figure.*

"It cuts my peripheral vision." *Shouldn't. Not if it's fitted properly. Besides, you need to be turning that head and keeping your scan going anyway, not trying to peek out of the corners of your eyes.*

"It's hot." *Damn right. Feels good in the winter. Maybe not so good in summer, but next time look for a vented model.*

"It's heavy. Makes my neck tired." *It won't, if you keep wearing it. Your neck muscles probably need the conditioning, or they wouldn't be getting tired.*

"I just don't like 'em." *Well, there ya go. Probably the only truthful and honest answer in the bunch. But ya gotta ask yourself, "Have I lived all I want to? Do I want to see my kids graduate from school? Do I want to live to enjoy my grandkids? Buy a good helmet-and wear it.*

BOOTS OR SHOES? PROTECTION VS STYLE—"NASTY FEET" AND FLIP-FLOPS.

It seems weird, but the same guys who won't wear a helmet often are seen wearing the big, klunky engineer boots. You know, the square-toed jobs with the straps and chains all over them? Those may look phony, but they offer darn good protection for feet and ankles.

I didn't mention it in the apparel section, but I'm always running across the guys who have the biker wallet on a chain, the bandanna, the jacket, the boots and the tattoo, but no bike. When you politely ask, "What kind of bike do you ride?", you'll get "Well, I'm *between* bikes right now…" Yeah, between your macho attitude and your wildest dreams, my man. Get a life. (As I watch him drive away in his pickup truck (cage) with the Harley sticker on the back window.)

These are the guys who want the look, but don't want to pay their dues. They'll never be caught in a hailstorm on top of the ridge between Cañon City and Cripple Creek, Colorado, with the lightning hitting the trees all around . . . but that's another story.

When selecting boots, I look for a non-absorbent sole, such as the Vibram hiking sole, full leather uppers and a tall shank or high top that goes up over the ankle and halfway up the calf. Whether you prefer lace-up boots or a slip-on type really doesn't matter all that much. Avoid slick soles such as the smooth leather used in the typical western boot. Save those for the square dance. When you're holding up an eight-hundred pound bike with your old lady and all your camping gear, and you happen to place your foot in doggie poop, you'll appreciate the extra traction. The more rigid the construction of the heel and sole, the more it will protect against foot injuries when you take that spill. Typically, any boot that is comfortable to walk in for any appreciable distance won't be any good for riding, and vice-versa.

Never ride in low shoes and never, *ever* ride in flip-flops. There again, it goes with the attitude. If you dress like you're going to be in a wreck, you may never have it happen. If you have a blasé attitude about riding, it will get you when you least expect it.

Remember, **there are three kinds of motorcyclists:** Those who have crashed, those who are about to crash, and those who are about to crash *again.* Presumably because they didn't learn a damned thing the first time.

ARE TATTOOS MANDATORY? *SKIN ART AND ITS POPULARITY.*

Here's one that'll floor ya. I happen to like tattoos. I have a bunch and I'm planning the next one. My wife tells our friends it's part of my mid-life crisis and I got them because I couldn't afford a Corvette.

Skin art, as you may have noticed, is enjoying an unprecedented period of popularity. Seems like everybody's getting tats and piercings these days. If you decide to go for the tattoo, here are some things you should know.

1. First of all, <u>yes, it will hurt</u>. About like a cigarette burn or a bee sting. I find that with a good artist, by the time it starts to really ache in one spot, he's moved on to somewhere else.

2. Don't do it on a whim. (Or when you're drunk or stoned) This is something that you and everyone else will see for the rest of your life, unless you're really into *serious* pain and want to try and have it removed. This can be done by

abrasion or laser, but it *is* a big deal. Figure the medical procedures for removing a tattoo will cost roughly ten times what the skin art did in the first place.

3. Most people find they enjoy their tattoos more if they have some meaning. It doesn't have to be a deep religious deal, but a lot of themes are overworked. You need something different, that's really "you". Hint: I did all mine on my computer using clip art and photos before I ever went to the artist. Selecting something from the "flash" he has plastered all over his walls may suit you fine, but if you want something special, figure on creating it yourself or having an artist work it up for you.

4. Once you get it, take care of it. It will need to be coated with Neosporin the first day and A&D Ointment for the next three or four days. Then it will peel, about like a sunburn. After the peeling, you're good to go, but remember, sunlight will fade the colors. If you tan,

protect it with sunscreen. Don't pick at it and you should keep a moisturizing lotion on it for a couple of weeks.

5. If you have a job where they might frown on having tats, do what I did: Hide 'em. When I'm in uniform, nobody would ever dream I have Jiminy Cricket on my shoulder or a big dragon on my back. When I want to show 'em off, I just wear a tank top.

6. I'm not into piercing, except for one earring, so you'll have to get info on that somewhere else. My earring took about five seconds at Claire's Boutique and was totally painless. I wear a diamond stud when I'm at work, and so far, it's caused no problems for me. There again, cleanliness is the key. If a piercing gets infected it can be a major deal. Especially in certain areas, such as on or around the genitals.

7. Most tattoo artists nowadays are very careful about sterilizing equipment and exposure to blood-

borne pathogens. Don't let just anyone tattoo you. Go to a licensed, inspected shop and don't be afraid to ask questions. Most artists are understandably proud of their work and their chosen profession. The *last* thing they want coming out of their shop is an infection. If you don't see sterilizing equipment, an autoclave and latex gloves in use, leave and go elsewhere. If you get attitude or the dude's hung over or stoned, same deal, go elsewhere.

BIKER VS MOTORCYCLIST WHAT DO YOU "WANNA BE"?

Okay, you got the bike, you got the gear, you got the training and you got your head on straight. Now, what's it gonna be? Are you gonna be a "weekend warrior", one of those who ride on Saturday or Sunday and let the bike set all week, in favor of the cage? Or do you choose to do something finer? Do you wanna make the motorcycle and all that it entails your way of life?

Personally, I choose to ride as often as I can. I've heard people say that the more you're exposed to the dangers of traffic, the more likely you are to become a statistic. I think that's just exactly ass backward. I think it's the rider who rides every day and deals with the traffic every day who soon becomes the safest, because he never loses his edge. He never loses sight of the fact he can be killed at any moment and he never relaxes entirely into the safety of the cage.

By now you've noticed I often refer to cars as "cages", and that's just what they are for me. More years ago than I care to admit or even think about, I immersed myself in the motorcycle experience, to a depth most people never achieve and it has made some indelible marks on my psyche.

To me a car or pickup *is* a cage. An enclosure that, while reasonably mobile and without doubt comfortable, I find confining and deceptive in that it gives a false sense of security to most drivers. Ever notice how brave people get when they're in their cages? It's like: "*This is my territory. I can do anything I want in here. Rain, sleet, snow and dirt can't get me, so you can't either, buddy. I can flip you off, be as rude as I want, talk on my cell phone and be stupid. In fact, I'm invisible in here. I can pick my nose, put on makeup, fondle my girlfriend in the most lascivious way and nobody can see me.*"

Actually, nothing could be farther from the truth. We *can* see you and you *are* vulnerable. I've been to enough drive-by shootings to know that cars are no protection when it comes to bullets. A crowbar will come right through that glass and you can be yanked out of there and into the real world in a heartbeat. A shot of mace or tear gas into the heater intake just in front of your windshield and in seconds, you'll come out all by yourself.

And I've seen far more deaths in car wrecks than on motorcycles. Car drivers will drive right into situations a biker would avoid like the plague. Nothing can hurt them, you see. Then they wind up crushed and mangled in the remains of their cages, being cut out and extricated with the Jaws of Life. If they'd felt just a little more vulnerable in the first place, maybe they would have been just a little more careful.

But in our minds, I think, we perceive our cars, with their leather seats and carpet and cup holders and air conditioning, to be extensions of our homes and your home *is* your castle, right?

Now think how much lighter the traffic would be and how much more courteous and careful everyone would be if the *only* vehicles allowed on America's roadways were motorcycles. If *nobody* had the "protection" and false security of their cage? Blows your mind, huh?

One thing constant motorcycling will do for you: It will make you a better driver, whether you're on the bike or in the cage. You'll find you'll be more courteous and you'll pay attention to all traffic situations, plus you'll darn sure look for motorcycles.

CLUBS, "GANGS" AND ROAD SERVICE. *GETTING STRANDED AND WHAT TO DO ABOUT IT.*

At times I have "belonged" to a number of motorcycle clubs. I have never been a member of a motorcycle "gang", and I'm not sure there really are such things. I think the motorcycle gang is just another Hollywood invention. I've found that even the worst-looking, scruffy, tattooed biker responds to a courteous attitude at least as well or better than most cage drivers.

Of course, when I'm decked out in biker gear, I always catch the random thoughtless remark about "Hell's Angels", just as, when I'm in security uniform, with a .40 caliber Glock strapped on, I catch the equally thoughtless remarks about, "Ya shot anybody lately?" So stupid. So uncalled for. But this is a free speech country, right? People never think about the officer's feelings when they pop off like that. They never think about the possibility that perhaps he has actually *been* placed in the position of having to take a life in a line-of-duty shooting. That maybe he had to make some nice lady a widow and take away someone's daddy because the man was stupid… well, again, that's another story.

At one time I belonged to the American Motorcyclist Association (AMA), the Gold Wing Road Rider's Association (GWRRA) and the Wichita Road Riders, all at the same time. I paid my dues and wore the patches and pins, went to the rallies and rides and poker runs and generally associated with the folks.

The police motorcycle experience gradually started me in the direction of getting away from all of that. It brought to my attention what rank amateurs most people are when it comes to actually riding their machines and the phoniness of many clubs and associations. The politics involved in most clubs finished the job. I learned over the years how little difference there is between the word "biker" and the word "bicker". I would say as far as clubs were concerned, I achieved burn-out for a number of years and spent my time riding alone. Recently, I have discovered the American Legion Riders and the Patriot Guard and have joined both. I still get nervous riding with a group and I tend to put myself at the tail-end, so whenever the wreck happens, I can see it and not be in it.

So, what do you do when you break down? Well, there used to be a universal, unwritten code among bikers that you never left another biker stranded without helping any way you could. Used to be, you could just set your helmet on the side of the road and that was the sign you were in trouble. I've stopped and helped others. I've carried messages, sent back trucks from dealers, and loaned money to folks I knew I'd never see again. Once I used a Harley as a tow truck and pulled a Suzuki Rotary a mile and a half back to a service area. Guy didn't even say "thanks".

Once, fifty miles east of Indianapolis, my bike died just as we hit the exit ramp to stop for gas. My wife and I pushed it a good hundred yards to the nearest filling station, where we were immediately joined by a man in a Chevy van, along with his wife.

After determining that my battery had died, he took us back to the city in his van, a one hundred mile round trip, to get a battery and see us safely down the road. He spent his Saturday afternoon helping us and wouldn't take a penny for his trouble. But now, with all the doctors, lawyers and yuppie wanna-be's that are out and about on bikes, I just don't know.

With some bikes, you get a membership offer in their exclusive club. Harley Davidson has "HOG" (Harley Owner's Group), Honda has Red Rider, etc. Most of these offer very reasonable road service policies. For a yearly fee, you can call their 800 number if you break down and they'll get you to a dealer somehow. For no more than they charge, it's a good idea, plus you get the patch, pin, decal and magazine, if you're into those things.

I do everything I can to be sure I don't break down, and then I pack some tools and a gun. Illegal? Yeah, I know, at least for most of us. But I'd rather have the good guys catch me with it, than the bad guys catch me without. Just don't *ever* try to take a gun into Mexico or Canada. They are not merely extensions of your own country. They will put you in jail and throw away the keys. The U.S. Constitution stops right at the border and the Bill of Rights, too. Better believe it, Buddy.

TOOLS FOR YOUR RIDE.
WHAT TO CARRY AND WHAT WILL WASTE YOUR TIME.

Part of my philosophy is the concept of Murphy's Law. "Anything that *can* go wrong, *will* go wrong and it will do so at the worst possible moment." Added to that should be, "Murphy was an optimist."

Counteracting Murphy's Law is the concept of the Guardian Angel. How could a crusty, pragmatic old ex-cop biker believe in a Guardian Angel? Simple. I've heard her voice, on more than one occasion.

At least a dozen times in my police career I heard The Voice say, "Watch it," or, "Don't go there," and moments later, the drunk sailed through the red light, or the old lady turned from the wrong lane. They were meant to get me, but the Angel saved my ass. What for, I don't know. Maybe so someday I could write this book. I hope that *wasn't* the reason, because if it was, then she's off duty now, and I'm probably screwed.

But this chapter is about tools. What should the biker carry? What is a waste of time? Unless you want to have a support vehicle follow you everywhere you go, (a practice that some groups actually engage in, the truck hauling the tools and beer and broken down bikes) you need a small inventory of tools. Most bikes come with a rinky-dink tool set of some kind. Some bikes actually come with a good set of tools. But, by and large, if you break down, you're gonna want something more serious than the little "Bag-O-Tools" you get with most bikes. At the same time, tools take up room and they're heavy. Pack what you *do* take at the bottom of your saddlebags, to keep your center of gravity as low as possible.

I carry a small socket set that fits flat in the bottom of my saddlebag. Bought it for $14.95 at Wal Mart. These are not Craftsman tools, or Snap-On. These are cheap, throw-away sockets that will get me by in an emergency, nothing more. In addition, I still retain the tool set that came with the bike. Add in a set of needle-nosed pliers with wire cutter feature, a good selection of fuses, about six feet of insulated wire and a test lamp. You can make your own test lamp from a bulb, two alligator clips and some wire leads, soldered directly onto the contacts on the bulb. Again, this is nothing fancy, just a quick way to see if connections are "hot". I also pack one extra spark plug, although I don't know why. I've never had a spark plug fail yet.

Remember when I said my wire leads for my battery maintainer were permanently attached to my battery? I also have a twelve-volt "drop light" that plugs into those same leads. Equipped with twelve feet of cord, it will reach anywhere around the bike and give good, strong light for several hours without draining the battery too much. Why not just use a flashlight? Because I tend to forget it's there and when I need it, the batteries are dead and corroded into powdery lumps of useless white stuff.

I don't carry a tire patching kit, though I know guys who do. I don't trust patched tires enough to ride on one. I'd rather call the tow truck and buy a new tire. If you have tubeless tires, you might consider a can of that "Fix-a-flat" stuff. It's basically compressed air and sealant foam in a can, with a screw-on device to fit the tire valve. I carried that stuff for thirteen years when I had my '84 Gold Wing and never needed it once. Talk about cheap insurance…If you have tube-type tires, don't waste your money, cause it won't work. Throw in a small flashlight to help you see or signal for help if you don't like my drop light idea and that's about it.

Your best insurance against a breakdown is still the maintenance you do *before* the trip. I seldom have trouble on the road, but then, like I said, I'm anal about maintenance. I have an acquaintance who blew a tire at Sturgis and crashed because he never looks at his tires. He left for a road trip with a back tire that was already bald and wore it *through the cord* until there was nothing there. Murphy's Law kicked in, wrecked his bike and broke his arm. I sold him that back tire new, when I was working at the dealership and I think he blames me for the accident, but I'm not sure. Doesn't matter, he screwed up and it bit him in the ass. Sometimes, even Guardian Angels get disgusted and just let shit happen.

BOOZE, DRUGS AND THE DRT CYCLIST *(DEAD RIGHT THERE)*

Police Officers, Emergency Services people and Firefighters develop a black sense of humor. Either that, or they don't last long and soon move on to other, less stressful occupations. It's merely a coping mechanism to help themselves deal with what they see every day. When people see this displayed on a scene somewhere, they feel we're the coldest, most callous and disrespectful bastards in the world, but it's just not so. Sometimes, though, if you can't laugh, you'll end up crying, and you can't afford that. The job comes first. If you need to cry, you do that on your own time.

Hollywood likes the term, "DOA" for Dead On Arrival. Once in a while you'll also hear, "DOS" for Dead On the Scene. Cops use the term "Code Black" which is the worst of the color codes. Cop humorists like the term "DRT" for "Dead Right There." They reserve this for the unfortunate few who are the victims of the sudden, irreversible, cataclysmic accident, when death is so sudden the victim's face may actually register surprise, rather than pain.

It's reserved for idiots who remove themselves from the gene pool by doing incredibly stupid things, like the guy who made the Darwin Awards one year when he was crushed by a 400 lb. Pepsi machine he was rocking, trying to get a free pop. Or another I read about who decided to bungee jump using those tie-down straps you use for hooking things to your bike. He duct taped a bunch of them together and jumped off a bridge. Only trouble was, he had more cord than the distance from the bridge to the roadway 75 feet below. Massive trauma. DRT.

Cycle riders who drink and drive are usually fitted neatly into this same category. We've already spoken about the skills required to operate a motorcycle and the rider's vulnerability in traffic. So why would anybody deliberately drink something that affects balance, coordination, depth perception, reaction time and judgment, then get on a bike? Beats me. But they do and the result is devastating. Even more incredibly stupid is the guy who thinks a little cocaine will make him into Kenny Roberts.

Dealers in the U.S. can sell you a

powerful, high-performance motorcycle with no credentials required. You'll never have to prove you can ride or show any particular skill level. The driver's licensing validation process for a cycle operator's permit ranges from a mildly effective riding test in some states to a grotesque joke in some others.

The trouble is, the joke's on the rider. There are probably fewer people in the world that could operate, say, a Kawasaki ZX14 Ninja to its full potential than there are qualified stunt pilots. Yet as soon as some kid gets a snootful of liquor or drugs, suddenly he can do anything with that machine.

The worst part for the officer is not working the scene, or going to the morgue to view and tag the body. It's the late-night visit to inform the family that their son was stupid and now he's dead. Their hopes and dreams may have been cooling on a slab for hours before they are located and told, especially when the driver carries no ID. So, when you ride, ride straight. It's easier on your family, and also on the officers and EMS people.

Have I ever taken a drink and then

gotten on a motorcycle? Yep. *A drink. One.* That's my limit if I'm on the bike, usually a beer with my Mexican food. I can have two if I'm driving the car. I can have all I want if *you're* driving…

Telling the cop who stops you that you've just had two beers isn't going to cut it, by the way, when you're totally wiped. Any cop that's been through DUI enforcement school knows a drunk when he sees one and uses a number of ways to check your balance, coordination, reaction time and your ability to comprehend and follow instructions. He actually will be using a chart to log the results and before he ever has you blow into the machine he will be able to predict within a few percentage points what the result will be. He already knows you're going to jail. That's why he orders a tow truck for your bike…

In most places nowadays, DUI is no longer a minor infraction and very few cops take drunk drivers home anymore. The old "slap on the wrist" for first-time offenders isn't happening, either. Figure the first DUI will run about $1500.00 to $2000.00 and get your license restricted. From there it gets worse very quickly, ending in most jurisdictions with loss of license and a good shot at felony prison time. That's right, locked away with serious and violent offenders for a while.

In spite of all that, I see all kinds of bikes parked at bars all the time and I'm sure those riders aren't in there sipping cokes or lattes.

YOUR ID STASH
DOCUMENTS THAT CAN SAVE YOUR LIFE.

On the side of my helmet I have taken the time to affix two strips of red embossing tape, the kind you use with a "Dymo" label maker. They say, "K. J. Crist", and "A Pos-NKA".

The first, my initials and last name, indicate that I'm *not* wearing someone else's helmet. The second line indicates my blood type, A Positive, and that I have No Known Allergies-NKA in medical terms, recognizable by any EMT, paramedic or ER doc who's worth his salt.

In my wallet, along with my driver's license, is a small packet containing emergency information—whom to contact in the event of injury or my demise, the fact that I'm an organ donor, and the papers that can get my remains shipped home from anywhere in the world at no cost.

This was fairly simple to accomplish. I merely typed what I wanted the authorities to know into my computer and printed a copy, signed it before a notary and it became a legal document. It looks sort of like this:

Attention

The bearer, _______(me)_________________, is an organ donor. Blood type is "A" positive, with no known drug allergies. In the event of death or injury, contact _________(wife)_________________ at _______(telephone)_____, or _______(daughter)_______________ at _________(telephone)_________________ as soon as possible. Permission is hereby given to use any and all means to sustain life until contact with _______(wife)________________ is made, after which she will make decisions. In the event of death, harvest any viable organs and contact the above person(s). My insurance carrier is ______(company)_______________, policy number ______________________________. My vehicle is to be impounded and held for the above person or my insurance carrier. In the event of my death, Resthaven Mortuary of Wichita, KS guarantees shipping from any point in the world.

This is notarized to show the date it was signed and the fact that I did, in fact, sign it. If the local cops and medical personnel can't figure out what to do after reading all this, I probably shouldn't have been riding in a part of the world where people are that backward, anyway.

IF IT'S SO DANGEROUS ... (WHY WOULD ANYBODY DO IT?) THE LOVE OF MOTORCYCLING.

I think the reason most people don't last long with motorcycling or adopt the "biker lifestyle" is because they never get serious enough with it in the first place.

People like to dabble around and try new and exiting things. Bungee jumping. Hang gliding. Extra-marital sex. Whatever. They buy a bike, ride a while, maybe get hurt once and then it's "been there-done that" and on to whatever comes next.

We live in a country where we are generally affluent enough to enjoy our hobbies without cutting into our livelihood money.

I was fortunate enough to have friends who went into the biker thing with me. A few don't ride any more and I outlived the rest. I stuck with it because it suits me.

I like the wind, the noise, the vibration and the challenge of staying alive year after year on a bike. I like taking trips to places most of my friends only talk about going to "someday".

I like seeing guys who are also on vacation, but stuck with a car full of screaming kids and a fat old lady, drooling when I go by, wishing to God *they* could be with *me*. I even enjoy the war stories you always get from the old timers who hang out at gas stations and tell you about their Harley or Indian they had back in the forties.

I like the admiring glances I get from kids and women and the comments about the "pretty bike".

A big, fancy road bike is a real conversation starter for a lot of people, those who tend to be outgoing and honest. For the rest, it tends to intimidate and they leave you alone. I've probably met a thousand people in my life, not counting those I met when I was a cop, whom I never would have met, were it not for the bike.

Would I be safer in a cage? Marginally. Would I stay cleaner? For sure, but I'm not afraid of soap and water. What would I miss by not being a biker?

I'd miss the smell of the pine forest in Maine, so strong it almost intoxicated. I'd miss the thrill of Colorado mountain roads, where the curves are serious and the guardrails are made out of fresh air. I'd miss the silence of the New Mexico desert, where you can stop and your ears will ring with city sound for fifteen minutes before you can hear the wind and the insects moving around.

And now that I've finally reached retirement, I tend to ride more than ever, in spite of my age. I've said it often and I'll say it again: The *worst* ride I ever took on a motorcycle was still better than the *best* ride I ever took in a car. No question.

CHICKS AND BIKES
'NUFF SAID.

There is a certain kind of woman out there who is attracted to motorcycles and the guys who ride them. She may be pretty, or maybe not. She may crave danger or the attention bikers get. Maybe she likes drugs and thinks bikers have better dope. Maybe she has a death-wish or just wants to piss off her parents, boyfriend or minister. Maybe there's some other guy lurking in the wings, waiting to "catch" her out with you, so he can beat your ass.

The point is, you want to be careful. Especially if you're married. If you're single, maybe you'll *want* the adventure of dating a gal that just got interested because you were on a bike. I wouldn't recommend it.

I think the best way to meet women is *without* the bike. Let them find out in due time and you can gauge their reaction to it. That can be quite revealing, too.

You say things were going well until she saw your motorcycle, and suddenly you're

Typhoid Mary? If she's that close-minded, it probably wouldn't have worked for long, anyway. If you're seriously into bikes, don't let some chick con you into marriage until it's understood that you *won't* sell the bike just because she got pregnant.

The union of two people is supposed to be about mutual happiness and it won't work if one of them is miserable. If you're walking on your lips every time a scoot goes by, the union won't endure. My old lady doesn't ride much any more, but she knows it's important to me, so she leaves it alone.

I don't sleep well at night if there's not a bike in the garage. It's not that I actually go out for late-night rides that often any more, but it's just knowing I *can*, if I want to. It's there whenever I'm ready.

STAYING IN SHAPE
PHYSICAL FITNESS AND THE SERIOUS RIDER.

Okay, here we go. This is where it gets weird. The part where I tell you how pumping iron and running will make you better suited to ride. But you really know that riding a bike's just sitting on your ass and playing with some controls, right?

Wrong again, my friend. I'm now seventy-five years old. I start every day with a four-mile walk and I pay attention to what I put into my body. When I was younger, I hit the gym every day and could bench press my own body weight plus fifty pounds any old time.

I have come to believe that the better shape you're in physically, the better you will feel on the long rides and the quicker your *mind* will be when it counts. Like when some cage driver tries to smear you onto the pavement.

I quit smoking after 21 years, when I was thirty-eight and I can sympathize with anyone who's trying to quit. It's one of the toughest addictions there is, and if you can't kick it, my advice is: Get help! Every day that you suck smoke from burning leaves into your lungs is a day without sunshine, man.

I still drink an occasional beer, and I see nothing wrong with that. I never drink more than two if I'm driving, or one if I'm on the bike. They say alcohol kills brain cells, but I think I'll probably have enough to last me.

But even if you're not into fitness, some conditioning exercises can't hurt, and on the long rides, when you get off for a stretch, you might try really stretching, like you would before a run. It gets the blood flowing and helps clear the mind.

DEALING WITH INJURY
SOONER OR LATER YOU WILL GET HURT.

We all get hurt from time to time. The German philosopher Friedrich Nietzche said, "That which does not kill me, makes me stronger." Pain is also nature's way of letting you know you didn't kill yourself *that* time.

In this chapter, I'm not talking about catastrophic injury, like from a crash. If that happens to you, chances are you won't have much say in what happens from the time the Emergency Services unit arrives. I've been extremely lucky. I've never lain on the pavement, hearing the sirens coming and knowing they were for *me*.

What I want to discuss is the small stuff. The burn from the exhaust pipe. The bee sting. The "unimportant" road rash.

You should pack a first aid kit for any road trip. Pack as big as you have room for. If you never need it, thank whatever deity you believe in the next time you say your prayers and get on with your life. But if you need it, and you didn't bring it…oops!

You can make up your own kit for a few bucks and here's a list of what I would include:

Band-Aids-Take one box and dump it out, then take sizes from several boxes and put in just a couple of everything—you can always stop at a store and buy more of what you need.

Bee sting kit-Find a product called "Sting-Ease" if you can. It's usually on a card and consists of small glass ampoules that you break and the medicine soaks a piece of cotton. You then dab it on the sting. It actually works to relieve the pain. If you're allergic, that's another story. That would be information for that ID stash we talked about earlier, and they make a treatment kit for that, too. On a road trip in Montana I once ran through a swarm of hornets or Yellow Jackets. I was the lead bike and I didn't get a single sting. Everybody behind me did, though, multiple times. Out came the bee-sting kit and everyone was glad I had it.

Q-tips, cotton balls, tweezers, nail clipper, small scissors, etc. in a waterproof bag.

Aspirin, Tylenol, or whatever you like to use for headaches and pain.

Burn and antiseptic ointments. Small tubes will do-just enough for initial treatment, again, you can stop and buy more.

Roll of gauze.

One Ace bandage. A stick or newspaper or magazine can make a passable splint, when held in place with an Ace bandage. Also they're good to keep large dressings in place.

About four large **gauze pads**.

Some kind of **eyewash**—I prefer the most natural kind, rather than Murine or Visene.

Prescription Medicine-I prefer to keep this separate, and it should also be listed on your emergency information sheet for medical personnel to find. Keep in mind that if you go into a foreign country, your prescription meds must always be in the original bottle with the label attached, otherwise they can be deemed contraband and you could be busted for "smuggling" drugs.

TRAILERS AND SIDECARS
DO YOU REALLY WANT TO DO THAT?

No. You really don't. Now that I've said that and pissed off all the trailer people, let me clarify *why* I said it. Pulling a trailer would never work for me. Why? Well, first of all, I don't need it. I don't camp and I find I can carry everything I need on the bike, with a little jamming, cramming and other constructive packing methods. Second, I ride way too fast to be pulling anything behind me that has no brakes of its own. I have yet to see a braking system for a small trailer that worked. They either lock the wheels or they do nothing effective. And third, I have seen too much of the results of pulling trailers behind two-wheeled vehicles.

If you think pulling a trailer would work for you, first, analyze your riding style. Are you content to keep your speed down to levels where you might have a chance if something goes wrong?

Remember, anything bad that happens while you're pulling a trailer is going to be made that much *worse* bcause it's back there and it's hooked to the ass-end of your machine. When you brake sharply, the rear wheel becomes light anyway, because of weight transfer. With a trailer attached, it will be trying to pass you and most likely jackknife. In hard corners, it will be trying to push your back wheel out from under you as it continues to want to go straight, due to its own inertia.

Just like so many other things I see all the time, I shudder whenever I see a motorcycle go by with a trailer attached and I get as far away from them as the traffic will allow. Not because I think they're going to get me into an accident. I just don't want to take time to stop and fill out a witness report whenever they crash. The bigger the trailer, the worse I hate to see it. I've seen guys pulling a *boat* with a motorcycle. Sometimes I think a psychiatric exam should be mandatory to obtain a motorcycle operator's validation.

At any rate, if you choose to pull a trailer, make sure you have good insurance and I *would* buy the extra medical coverage.

At the Aspencade Motorcyclist Convention in Ruidoso, New Mexico, must have been in about 1979 or 1980, I met a guy named John. I have lost his last name, but he had a dog named Beano, who had his own trailer. Beano was a big collie or shepherd mix and an expert at the motorcycle game called the "Weenie bite", in which a hot dog is dangled above the level of the rider and the passenger must lunge upward and bite off a hunk. It's usually liberally slathered with mustard just to make it interesting.

After everyone else had their turn, John would sit the dog on the back seat and drive his big BMW slowly under the weenie and Beano would take the weenie, string and sometimes the flimsy wooden rig that it hung from.

John was also the guy who got pissed off one year at the phoniness of the "Best Dressed" competition. Folks were showing up in outlandish outfits that you couldn't possibly ride in, so he went downtown and appeared shortly in a tuxedo, complete with tails, cummerbund, top hat, spats and a walking stick. I don't remember if he won, but he certainly showed everyone how ridiculous the competition was getting.

John also showed me the only legitimate use for a trailer pulled behind a motorcycle. You see, Beano was John's companion. I think they actually lived on the road and John was apparently willing to take the added risk to have Beano along. I believe I'd have bought Beano a sidecar.

I had a Vetter Terraplane sidecar that I pulled with a 1978 Gold Wing for two seasons, until my daughter got old enough to drive. My own experience with a sidecar proved several things to me. First, a sidecar can be a safe and fun solution to the "too many people or too much stuff" problems. Second, a sidecar can be an attractive nuisance and a great big pain in the ass. It completely changes all the rules about motorcycle riding. As soon as the sidecar is installed, forget about counter-steering. Now you're a three-wheeled vehicle, like a tricycle. Whichever way you point that front wheel, you're going there. Plus, whenever you accelerate, the cycle will pull right and when you brake, it will pull left because the sidecar is trying to pass you. Whenever you make a right turn, the sidecar will want to lift its tire free of the pavement, particularly when it's empty.

Whenever you park the damn thing, you'll be lucky if you don't have people crawling all over it and sitting in it when you return. People who'd never *dream* of breaking into your car will sit in your sidecar without permission and let their kids climb all over it.

You'll also find any number of cage drivers who'll want to drive beside you for mile after mile, admiring your rig. This will soon bother you, because they'll be spending more time staring at you than at the road. This tends to be dangerous. (Remember the "circle of safety"?)

On the other hand, if you're a real "people person", sidecars are a great conversation starter, and not a bad way to pick up chicks. Even old grannies will ride in a sidecar with a scruffy-looking biker. If you're into old grannies, this would definitely be the way to go…

The trade-offs are:

Lower fuel economy; Figure it costs about five miles per gallon to pull a sidecar.

Extra tire wear; The back tire will scrub off tread rather quickly due to the side-pull of the sidecar. If the toe-in on the outer wheel is not set correctly it can cut your tire life in *half*.

Extra brake wear; Stopping the extra

weight of the sidecar and contents will take its toll.

Reduced performance; Obviously, the extra weight and wind resistance will cut top speed and acceleration unless something is done to enhance performance.

Possible frame or other structural damage; Unless the bike was designed to take a sidecar, its frame probably will be stressed beyond intended design limits. This may or may not prove to be a problem. On my Gold Wing, I found the addition of a front fork brace was necessary to cut down on flexing of the fork tubes, and it would still wobble at low speeds. On most bikes, installation of a sidecar or trailer will void the warranty, so I wouldn't do it to a new bike.

"WHERE DO YOU STAY?" *THE "MOTEL CAMPER" AND WHY SOME MOTEL OWNERS HATE YOU.*

I never camp. Period. I tried it twice, once in my Harley riding days at Roaring River State Park (that's in Missouri) and once in a roadside park near Pittsburg, Kansas a few years later. Both times were disasters. I know guys who do it and swear it's the greatest. Doesn't work for me.

The first time, the damned air mattress kept deflating all night and the bugs ate me alive. When we got up in the morning everybody in the group was sitting around a picnic table eating cold cereal. My buddy and I looked at each other, cranked our old kick-start Harleys and went to the store. When we got back, we built a fire and had *real food.* Bacon and eggs, hash browns and biscuits, and real coffee.

The others, who didn't get to share in this bounty, asked, "Do you guys eat like this every morning?"

We said, "No, only when we've spent the whole night sleeping on friggin'*rocks.*"

The second time, I was trapped in a leaky tent, in the rain, with my wife and another couple. After the cold, wet tent fabric had slapped me in the face about a dozen times, I gave up trying to sleep and crawled out to have a cigarette. This was still in my smoking days. As I sat with my friend on a picnic table, there was a flash of lightning across the sky and there, standing out clearly against the black sky, was a perfect white tornado hanging from the clouds. The tent came down, we went to town and the motel room seemed cheap to us that night. I've been a motel camper ever since.

After a day in the heat, or rain, or cold, whatever—I want a good room with a door that locks, clean sheets, a good shower, and—what the hell, "magic fingers" in the bed won't hurt, either.

Trouble is, sometimes, as I'm turning into the lot at the Best Western or Travelodge, I notice the man flipping the sign to "No Vacancy". Could it be he really just sold the last room? Funny, there's only three cars in the parking lot. Perhaps he doesn't like bikers. Wonder why? And if I get a room, it will always be on the second floor. They *never* have a ground floor room. Coincidence? I doubt it.

Once, at the Aspencade Motorcyclist Convention in Ruidoso, New Mexico, a guy actually told me how he kept his Gold Wing so sparkling clean on road trips. This dipstick actually *bragged* about how, after they'd had their morning showers, he and his wife would *wipe down the bike* with the motel towels!

I said, "Then you're the dumb bastard who causes me to have trouble getting a room!"

Folks, whatever you do, don't be stupid enough to use the motel's nice white towels to clean your ride.

When you do that, they'll never be white again and the next rider to pull in there may be out of luck for a room. Don't put the bike in the room and drip oil on the carpets, even if it *will* fit through the door. Buy a cover, buy an alarm system, or whatever. Carry your own towels, if you're that anal about having a clean bike. Myself, when I'm on the road, I clean the lights, mirrors and windshield. The rest just has to stay dirty until I get home.

And, last of all, don't party and drink and raise hell all night and disturb everyone else. It makes it bad for all of us. I hate to hear the phrase, "Damn bikers kept us up all night."

WISDOM FROM OLD BIKERS.
STREET SURVIVAL–DRIVING FOR EVERYONE ELSE.

Among aircraft pilots there is a saying: "There are old pilots-There are bold pilots. There are no old, bold pilots."

The same is true of bikers. If you intend to survive, you need to apply wisdom to your riding. Jim Stansberry, my Police bike instructor, said that riding is 99 percent between the ears, and he wasn't talking about balance. *Take* the rider course. Take *every* course you can afford. I've often wished I'd taken the race track course when I was younger. Yes, there are such things. They're pricey, but you'll learn things you could never pick up on the street. The California Super Bike School comes immediately to mind.

Pay attention to everything that's going on around you. When I was a cop and we'd have to shut off traffic, I was always amazed at the number of people who would not see all the emergency vehicles and fire hoses and such until they were ten feet away. C'mon, people, you can see the red lights at least a half mile! Wake up and smell the exhaust fumes.

Control your temper. I had a lot of trouble with this at first, until I realized that I was just pointing myself at an early grave due to stress and heart attacks. I still mumble under my breath, but I don't let 'em *know* they scared me.

And last of all: Do it your own way. Don't try to be what ya ain't. Nothing is more disgusting than a guy with a thirty thousand dollar bike, two thousand dollars worth of riding togs, a magnetic earring and no riding skills. Kinda like walking into the kitchen at the Hyatt Regency and finding Chef Boy-Ar-Dee readin' recipes off cans.

Be careful out there and when ya see that "biker wave" always return it. You know the one. The rider points that gloved hand out and down at the ground? That is a constant reminder from each serious biker to every other serious biker to *keep the rubber side down!*

ABOUT THE AUTHOR

Kenneth James Crist retired from the Wichita, Kansas Police Department in 1992, after 20 years' service, the second half as a motorcycle patrolman. During his time on "the wheel", he was also a certified instructor for Motorcycle Safety Foundation, teaching others the basics of motorcycling and not getting killed in traffic.

He began writing in 1994 and has over 200 short stories to his credit in venues ranging from Skin and Bones to Kudzu Monthly, mostly internet publications.

Having retired a second time from the security force at Wesley Medical Center in Wichita, he now drives for American Red Cross blood services as a volunteer and is also active with the American Legion Riders and the Patriot Guard.

He lives with his wife Donna in Wichita, along with two spoiled dogs and a haunted Kawasaki, "Christine."

Made in the USA
Monee, IL
04 November 2026

67375820R00075